Suman Paul

Introduction to MANET and Clustering in MANET

Anchor Academic
Publishing

Paul, Suman: Introduction to MANET and Clustering in MANET, Hamburg, Anchor Academic Publishing 2016

Buch-ISBN: 978-3-95489-878-7
PDF-eBook-ISBN: 978-3-95489-873-2
Druck/Herstellung: Anchor Academic Publishing, Hamburg, 2016

Bibliografische Information der Deutschen Nationalbibliothek:
Die Deutsche Nationalbibliothek verzeichnet diese Publikation in der Deutschen Nationalbibliografie; detaillierte bibliografische Daten sind im Internet über http://dnb.d-nb.de abrufbar.

Bibliographical Information of the German National Library:
The German National Library lists this publication in the German National Bibliography. Detailed bibliographic data can be found at: http://dnb.d-nb.de

All rights reserved. This publication may not be reproduced, stored in a retrieval system or transmitted, in any form or by any means, electronic, mechanical, photocopying, recording or otherwise, without the prior permission of the publishers.

Das Werk einschließlich aller seiner Teile ist urheberrechtlich geschützt. Jede Verwertung außerhalb der Grenzen des Urheberrechtsgesetzes ist ohne Zustimmung des Verlages unzulässig und strafbar. Dies gilt insbesondere für Vervielfältigungen, Übersetzungen, Mikroverfilmungen und die Einspeicherung und Bearbeitung in elektronischen Systemen.

Die Wiedergabe von Gebrauchsnamen, Handelsnamen, Warenbezeichnungen usw. in diesem Werk berechtigt auch ohne besondere Kennzeichnung nicht zu der Annahme, dass solche Namen im Sinne der Warenzeichen- und Markenschutz-Gesetzgebung als frei zu betrachten wären und daher von jedermann benutzt werden dürften.

Die Informationen in diesem Werk wurden mit Sorgfalt erarbeitet. Dennoch können Fehler nicht vollständig ausgeschlossen werden und die Diplomica Verlag GmbH, die Autoren oder Übersetzer übernehmen keine juristische Verantwortung oder irgendeine Haftung für evtl. verbliebene fehlerhafte Angaben und deren Folgen.

Alle Rechte vorbehalten

© Anchor Academic Publishing, Imprint der Diplomica Verlag GmbH
Hermannstal 119k, 22119 Hamburg
http://www.diplomica-verlag.de, Hamburg 2016
Printed in Germany

Content

Chapter 1: Introduction: Mobile Ad Hoc Network (MANET) and Motivation 9
Overview 9
Motivation 11

Chapter 2. Concept Review 12
2.1 Mobile Ad Hoc Network Features 12
    2.1.1 Autonomous terminal. 12
    2.1.2 Distributed operation. 12
    2.1.3 Multi-hop routing. 12
    2.1.4 Dynamic network topology. 12
    2.1.5 Varying link capacity. 13
    2.1.6 Light-weight terminals. 13
    2.1.7 Bandwidth-constrained and variable link capacity. 13
    2.1.8 Energy Constrained Operation. 13
    2.1.9 Security Issue. 13

Chapter 3: Applications of MANET 14
3.1 Defense (Military battlefield applications) 14
3.2 Sensor Networks 14
3.3 Automotive Applications 15
3.4 Commercial Sector 15
3.5 Personal Area Network 15
3.6 Conferencing 15
3.7 Embedded Computing Application 16

Chapter 4. Issues and Challenges of MANET 17
4.1 Quality of service (QoS) 17
4.2 Routing 17
4.3 Spectrum allocation 18
4.4 Energy Efficiency 18
4.5 Security and Privacy 18

Chapter 5. Routing Protocols and Topology Management 19
5.1 Proactive Protocols 19
5.2 Reactive Protocols 20

| 5.3 | Hybrid Protocols | 20 |
| 5.4 | Topology Management in MANET | 21 |

Chapter 6: Clustering in MANET 22

6.1	Why Clustering?	22
6.2	Characteristics of Clustered Architecture	23
6.3	Clustering in MANET: Classification by properties, cluster-head capabilities, clustering process	24
6.4	Cost of clustering	27
6.4.1	Ripple Effect	28
6.4.2	Computation Round	29
6.5	Design parameters of clustering	30
6.5.1	Trust value	30
6.5.2	Degree	30
6.5.3	Battery Power	30
6.5.4	The Max (maximum) value:	30
6.5.5	Stability (also defined as Mobility)	30
6.6	How to maintain Structure of Cluster	31
6.6.1	Sending beaconing signal:	31

Chapter 7. Proposed Algorithm 33

Chapter 8: Clustering Simulation Framework for MANET 35

8.1	Architectural view of Network Simulator	35
8.2	Clustering Framework	35
8.2.1	Concepts	36
8.2.2	Steps of simulated algorithm	36
8.3	Simulation Result	37

Further Reading 39

Chapter 1:
Introduction: Mobile Ad Hoc Network (MANET) and Motivation

Overview

The emergence of powerful hand held devices like cell phones, personal digital assistants (PDA), pagers coupled with the advancement of wireless communication system have paved the way for a variety of mobile computing and wireless networking technology recently. The history of wireless networking goes to early days Defense Advanced Research Projects Agency packet radio network.

The advantage of wireless networking is its ability to support user mobility has created a new breadth to problem solving in this domain, resulting in unpredictable resource requirement and uncertainty in network connectivity. Solutions to these problems have boosted the market for wireless services.

Wireless ad hoc network is a collection of mobile devices forming a network without any supporting infrastructure or prior organization. Nodes in the network should be able to sense and discover with nearby nodes. Due to the limited transmission range of wireless network, multiple network "hops" may be needed for one node to another (source to destination or intermediate node) across the network. There are number of characteristics in wireless ad hoc networks, such as the dynamic network topology, limited bandwidth and energy constraint in the network. Mobile ad hoc network plays an important role in different applications such as military operation to provide communication between squads, emergency case in out-of-the-way places, medical control etc.

Routing protocols plays significant role in implementation of mobile ad hoc networks (MANET). Due to the characteristics of mobility of ad hoc networks it is crucial problem to find path or route from source to the destination node and perform the communication between nodes for a long period of time.

In MANET, a number of routing protocols using a variety of routing algorithms have been proposed. For example, Dynamic Source Routing (DSR), Ad hoc On demand Distance Vector Routing (AODV), Temporally Ordered Routing Algorithm (TORA),

Location Aided Routing (LAR) and periodic (proactive) protocols such as Destination Sequence Distance Vector (DSDV), Distributed Bellman Ford, where member nodes exchange routing information and knows a current route to each destination periodically. Also, several protocols uses both reactive and proactive mechanism such as, Zone Resolution Protocol (ZRP) Cluster Based Routing Protocol (CBRP) etc.

The basic idea of on-demand routing protocols, is that a source node sends a route request and makes routing decision based on received route reply. This route reply may be sent by destination or intermediate node. On-demand routing protocol has several advantages. It offers flexibility, correctness and simplicity. Although, an on-demand routing algorithm has the disadvantage of increasing per-packet overhead in the network. This overhead decreases the availability of bandwidth resulting increase of the transmission latency of each packet. As result, consumption of battery power is more in transmitter and receiver end. Due to flooding i.e. propagation route requests it is difficult to limit spreading of unnecessary packets.

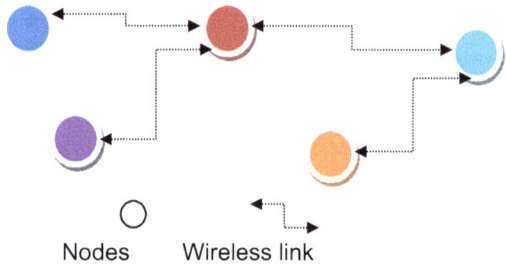

Figure 1: Infrastructure less (Ad Hoc) Network

The basic idea of proactive routing is periodically updating routing table via exchanging routing information. According to routing table, source node knows path or next hop to destination anytime when route needs. Route information is available as per need in proactive routing. This method creates delay prior to data transmission. However, proactive routing protocols are not suitable for mobile ad hoc networks, as they continuously use large section of the network capacity and bandwidth to keep the routing information. Proactive routing protocols tend to share out topological changes and alternation widely in the network. The creation of a new link or termination of a new link at one end of the network may not be important piece of information at the other end. Hybrid routing protocols have best characteristics of both reactive and proactive routing protocols. The main idea of the hybrid routing protocols is the limiting the set of forwarding nodes and using the proactive routing algorithm for nearly placed nodes which usually forward data to far placed nodes.

Motivation

In the last several decades research interest has increased in the area of mobile ad hoc networks. These are in contrast with wireless networks that depend upon a pre existing fixed infrastructure of base stations.

The main design goal of mobile ad hoc network was to meet the challenges as follows:

a) Military Application:

The situation in the battlefield requires soldiers to move from place to place without any constraint by wired communication and communication with each other is performed without depending on any fixed infrastructure. It is impossible to have a fixed backbone network in certain territories such as desert.

b) Propagation of electromagnetic signal:

The frequencies higher than 100 MHz are restricted by their propagation distance. Therefore, for communication between two nodes in a multi hop routing protocol is essential. This also justifies message must be transmitted from one node (source) to other node(destination) via intermediate nodes i.e. in hops.

Chapter 2:
Concept Review

2.1 Mobile Ad Hoc Network Features

A MANET has the following features:

2.1.1 Autonomous terminal.

In MANET, each mobile terminal nodes are autonomous and self-configuring nodes. These nodes perform as a host and a router. Besides the basic processing ability as a host, the mobile nodes can also perform functions of a router. So usually endpoints and switches are interchangeable and inseparable in MANET.

2.1.2 Distributed operation.

The control and administration of the network operations, are controlled and managed in a distributed type, among the terminal nodes. These nodes involved in a MANET should team up amongst themselves and each node acts as a relay as per requirement, to implement functions e.g. security and routing.

2.1.3 Multi-hop routing.

Ad hoc routing algorithms can be single-hop or multi-hop in nature. This categorization is mainly by considering different link layer attributes and routing protocols used in MANET. Single-hop MANET is simpler than multi-hop in terms of protocol architecture and for design and implementation. But there will be a compromise of lesser functionality and applicability. When delivering data packets from a source node to its destination node, the packets are forwarded via one or more intermediate nodes (in multiple hops).

2.1.4 Dynamic network topology.

As the nodes in MANET are mobile, the network topology varies rapidly and unpredictably with respect to time and space. As a result the link among the modes changes with respect to time. The MANET should become accustomed to the traffic and propagation conditions as well as the mobility patterns of the mobile network nodes.

The mobile nodes in the network dynamically establish routing among themselves as they move.

2.1.5 Varying link capacity.

High bit-error rates of wireless connection might be more in a MANET. One path may be shared by several sessions. The channel are prone to noise, fading, and interference. It has less bandwidth than a wired network. In some cases, the path between any pair of users can traverse multiple wireless links and the link themselves can be heterogeneous in nature.

2.1.6 Light-weight terminals.

In most cases, the MANET nodes are mobile devices with less processor capability, constraint in memory size, and low power back up. Such devices need optimized algorithms and mechanisms that implement the computing and also for communication.

2.1.7 Bandwidth-constrained and variable link capacity.

Wireless links have lower capacity than infrastructure wired counterparts.

2.1.8 Energy Constrained Operation.

Nodes in Mobile Ad Hoc Network rely on batteries as source for their energy.

2.1.9 Security Issue.

Mobile Wireless Networks are more prone to physical security threats than wired counterparts.

Chapter 3:
Applications of MANET

With the increase of use portable devices as well as progress in wireless communication, ad hoc networking is gaining importance. Ad hoc networking can be applied anywhere, where there is no communication infrastructure or the existing infrastructure is expensive or difficult to implement. Ad hoc networking allows the devices to maintain links to the network as well as easily adding as well as removing devices to and from the network. The set of applications for MANETs is diverse, ranging from large-scale, mobile, highly dynamic networks, to small, static networks that are constrained by power sources. Besides the inheritance applications that move from traditional infrastructure environment into the ad hoc context, a great deal of new services can and will be generated for the new environment. Typical applications include:

3.1 Defense (Military battlefield applications)

In modern battlefield requires robust and reliable (secure) communication military battlefield in many forms. Most communicating nodes are installed in mobile vehicles used for defense. Defense personnel could carry telecomm devices that could talk to a wireless base station or directly to other telecom devices if they are within the radio range. However these forms of communication are considered to be primitive. At times when wireless base station is destroyed by enemy, a soldier will be prohibited from communicating with other soldiers if the called party is not within the radio range. This is the scenario where mobile ad hoc networks come into play. Ad hoc networks are well known as self organizing networks since they are robust when nodes disappear due to destruction or mobility. Through multi hop communication soldiers can communicate to remote soldiers via data hoping and data forwarding from one radio device to another.

3.2 Sensor Networks

Another application of MANETs is sensor networks. This technology is a network composed of a very large number of small sensors. These can be used to detect any number of properties of an area. Examples include temperature, pressure, toxins, pollutions, etc. Applications are the measurement of ground humidity for agriculture, forecast of earthquakes. The capabilities of each sensor are very limited, and each must rely on others in order to forward data to a central computer. Individual sensors are limited in their computing capability and are prone to failure and loss. Mobile ad hoc sensor networks could be the key to future homeland security.

3.3 Automotive Applications

Automotive networks are widely discussed currently. Cars should be enabled to talk to the road, to traffic lights, and to each other, forming ad-hoc networks of various sizes. The network will provide the drivers with information about road conditions, congestions, and accident-ahead warnings, helping to optimize traffic flow.

3.4 Commercial Sector

Ad hoc network can be used in emergency/rescue operations for disaster relief efforts, e.g. in flood, fire, earthquake. Emergency rescue operations must take place in case of nonexistent and damaged communications infrastructure and rapid deployment of a communication network is required. Information is relayed from one rescue team member to another over a small handheld device with portability. Other commercial scenarios include e.g. ship to ship ad hoc communication, enforcement of law etc.

3.5 Personal Area Network

Personal Area Networks (PANs) are formed between various mobile and fixed nodes in an ad-hoc manner, e.g. for creating a small network. Nodes can form an autonomous network, interconnecting various devices. PANs becomes more meaningful when connected to a larger private or public network.

3.6 Conferencing

Mobile Conferencing one of the most recognized application. Set up of an Ad Hoc network is essential for mobile users where they need to gather in a project outside the office environment in an ad hoc manner.

3.7 Embedded Computing Application

Computing internetworking machines offer flexible and efficient ways of establishing communication methods with the help of ad hoc networking.

Chapter 4:
Issues and Challenges of MANET

4.1 Quality of service (QoS)

The parameters of performance level of services are known as QoS. Different applications have different QoS requirement. These QoS parameters may be computes on per link, per flow, per node basis. Delay, jitter, packer loss rate, bit error rate, bandwidth are some of the QoS parameters in MANET.

4.2 Routing

Routing is a crucial issue in MANET. The conventional routing protocols used on wired networks do not perform satisfactory in MANET, which has basic characteristics of mobility and rapid membership change. In Ad Hoc networks, we need new routing protocols for following reasons:

i) The mobility resulting change of topology of the network which may be dynamic in nature.

ii) Existing protocols show least desirable behavior when presented with a highly dynamic and frequent changing interconnection topology.

Existing routing protocols face heavy a computational load on each mobile computing device for the issues of processing power consumption and memory size.

Existing routing protocols are not designed for dynamic and self-starting behavior as required to utilize Ad-Hoc networks.

Existing routing protocols for example, Distance Vector Protocol take a lot of time for convergence upon the failure of a link, which is very frequent in Ad Hoc networks. Existing routing protocols suffer from looping problems either short lived or long lived. Methods adopted to solve the problem of looping in traditional routing protocols may not be applicable to Ad Hoc networks.

4.3 Spectrum allocation

The problem of device mobility, interference, limited range, limited data throughput and sharing of the Radio Frequency spectrum amongst devices are various crucial issues. Most experimental Ad hoc networks are based on the ISM band. To prevent the problem of interference, Ad hoc networks must operate over some form of specified spectrum range. Microwave ovens operate in 2.4 GHz ISM band, which may interfere with wireless LAN systems.

4.4 Energy Efficiency

Energy Efficiency is a concern in MANET. Most existing protocols do not pay attention of power consumption. However mobile devices today mostly operated by batteries. Battery technology is still lagging behind the microprocessor technology. The lifetime of a Li-on battery is very low.

4.5 Security and Privacy

Wireless links in an ad hoc network vulnerable to various link attacks like message replay, message distortion, passive eavesdropping, active impersonation. Eavesdropping may cause access to secret information, violating confidentiality. Active attacks may cause injection of erroneous messages, modification of messages, delete of messages and to impersonate a node. As a result, creating problem of integrity, authenticity and availability.

Chapter 5:
Routing Protocols and Topology Management

Since the arrival of DARPA packet routing networks in the early 1970s, numerous protocols have been contributed for ad hoc mobile networks. These include high power consumption, low bandwidth and high bit error rates. An Ad hoc protocol is a convention or standard that controls how nodes come to agree which way route packets between computing devices in a mobile ad-hoc network.

Routing protocols in MANETs can be classified as:

1. Proactive(Table driven)
2. Reactive (On demand)
3. Hybrid(Combining the features of both)

5.1 Proactive Protocols

Proactive protocols are also known as table- driven protocols .In these protocols, all nodes maintain routing information to every other node in the network in its routing tables which are periodically updated with the change of network topology. In these protocols the difference exists in the way the routing information is updated, detected and type of information kept at each routing.

Some of these protocols are:

- Destination Sequenced Distance Vectored (DSDV)
- Distributed Bellman- Ford (DBF)
- Wireless Routing Protocol (WRP)
- Cluster head Gateway Switch Routing (CGSR)
- Source Tree Adaptive Routing (STAR)
- Hazy Sighted Link State Routing ((HLSR)
- Hierarchical Stare Routing (HSR)

5.2 Reactive Protocols

Reactive protocols are called on demand (as per requirement) protocols. These routing protocols are designed to minimize the overhead by maintaining the information for active routes. Delay is a issue in these protocols. Routes are determined and maintained for communicating nodes which need send data to particular destination. Route discovery is implemented by flooding a route request through the network. This scheme is relevant for Ad hoc environment. since the battery Power of the network nodes is conserved both by not sending the route request advertisements and by not requiring to receive them (a node could otherwise minimize its power consumption by entering itself in a sleep mode or in a standby mode when they are not busy with other tasks).

Some of the protocols are:

- Associativity Based Routing (ABR)
- Dynamic Source Routing (DSR)
- Temporary Ordered Routing Algorithm (TORA)
- Adhoc on Demand routing protocol (AODV)
- Cluster Based Routing Protocol (CBRP)
- Relative Distance Micro discovery Adhoc Routing (RDMAR)
- Signal Stability Routing (SSR)
- Caching and Multipath Routing (CHAMP)
- Ant-based Routing Algorithm (ARA)

5.3 Hybrid Protocols

Hybrid combines the merits of both proactive and reactive routing protocols with some additional features. The main idea of the hybrid routing protocols is the limiting the set of forwarding nodes and using the proactive routing algorithm for nearly placed nodes which usually forward data to far placed nodes. While route to nearly placed nodes is available immediately, there is no misuse of bandwidth due to propagation of the local information to the far placed nodes. Also with the flexibility and accuracy of the reactive routing, the overhead is greatly minimized caused by limitation of number of forwarding nodes. Hybrid routing algorithm does not focus on the route maintenance against node mobility. Improper balance between proactive and reactive routing causes degradation of performance of data transmission resulting higher end to end delay and of course, reduction of a packet delivery ratio. Zone Resolution Protocol (ZRP) is an example of Hybrid protocol.

5.4 Topology Management in MANET

Since the nodes in Ad Hoc Networks are mobile, they are continuously moving in and out of transmission ranges of their neighbors. Thus the topology formed by the initial selection of cluster head cannot be maintained indefinitely. Since ordinary nodes communicate through their cluster head, the network needs at least minor adjustments any time a node moves outside the transmission of its cluster head .If a node has moved into the transmission range of another cluster head, re-affiliation can transfer the node from one cluster to other. The re-affiliation (detachment of a node from its current cluster and attachment to another existing cluster) must be propagated only to the node, and old and new cluster head involved. On the contrary if the node moves outside the transmission range of any of existing cluster heads or if the mobility of the cluster heads themselves causes an unacceptable reduction of network coverage, then the dominant set must be updated. The change in dominant set must be propagated to every cluster head in the network.

Communication and computation cots incur in updating the routing protocols for the new dominant set. Messages must be dropped during the updates.

The topology management requires both clustering and routing algorithms to work efficiently hand in hand to achieve the ultimate goal of maintaining stability of Mobile Ad Hoc Network. Thus the research in this field should be a trade off between marinating a nearly optical network configuration while avoiding the high overhead caused by frequent dominant set updates.

Chapter 6:
Clustering in MANET

Clustering is an emerging research topic in MANET. In the context of mobility and large number of mobile nodes, clustering provides the guaranteed system quality of service (QoS) and system performance such as throughput, delay

6.1 Why Clustering?

Clustering in MANET means the he concept of dividing the geographical region to be covered into small regions. A large number of nodes in a MANET are divided into smaller number virtual groups.

The main objective of clustering in MANET is:

To partition the nodes in a number of clusters in order to make system scalable, to minimize expensive long range message communication and to increase the availability to increase the services locally. In each cluster, basically there are three different types of nodes:

a) Cluster head:

Depending upon certain parameters or deciding factors a particular node is selected as cluster head (CH).This head is responsible for coordination of communication and maintenance of cluster.

b) Gateway node:

Another type of node known as gateway node (GW). The gateway node in each cluster manages communication with adjacent cluster. It is link between two or more number of clusters. For simplification and smooth processing their may be more than one gateway nodes in a particular cluster.

c) Member node: A node which belongs to a particular cluster.

Clusters may change dynamically their topology affecting the mobility of the underlying network. Due to the dynamic nature of the mobile nodes their association and dissociation to and from the cluster affect the stability of the network and thus the reconfiguration of the cluster head is unavoidable. This is an important issue since frequently cluster head changes adversely affect the performance of other protocol such as scheduling, routing and allocation of resource. There must be solutions for efficient ways of interconnecting the nodes such as the latency of the system is minimized and throughput is maximized. Most of the cases, for electing the cluster head do not make an optimal solution with respect to load balancing, battery usage and MAC functionality.

A good clustering scheme should have the graph structure as much as possible when nodes are moving and as well as the topology is changing. Otherwise re-computation of cluster head and regular information exchange among the participating nodes will result high communication and computation overhead. Thus the optimal selection of cluster head and partitioning of nodes into clusters are essential characteristics of MANET.

6.2 Characteristics of Clustered Architecture

A clustering is a similar with single hop cellular architecture. There are four major characteristics of clustering architecture. First, there is only one cluster head in each cluster. Each node in a clustering architecture is either a cluster head or adjacent to one or more cluster heads. Any two cluster heads are not adjacent to each other. Any two nodes in the same cluster are at most two hops away from each other. Figure 2 illustrates the clustering architecture for an ad hoc network.

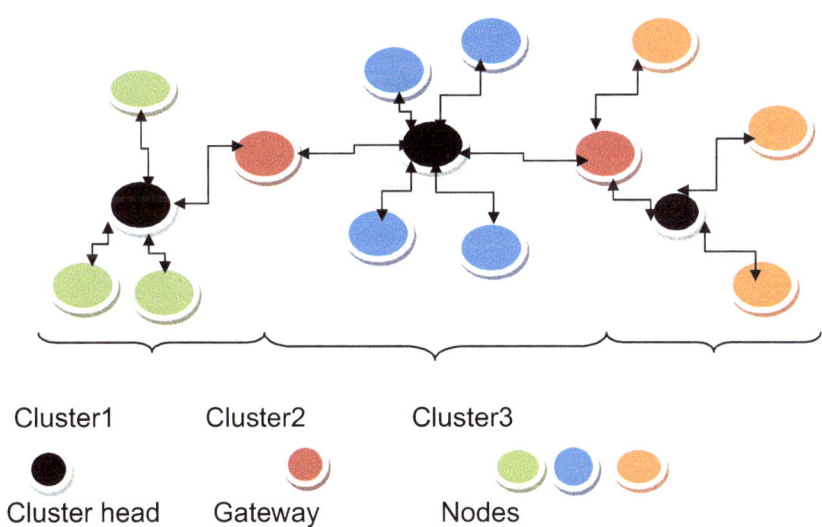

Figure 2: Simple clustered architecture in MANET

There is a link between two nodes if the nodes are within the transmission range of each other. In the figure the black nodes denotes the Head of the clustering architecture.

Clustering architectures can be categorized into two categories overlapping and non-overlapping. In overlapping clustering architecture a node which is not a Head to any of the cluster may belong more than one clusters and these nodes are known as gateway node. Communication between any two adjacent nodes have to rely on these common gateway nodes.

In the figure 2 there is no common gateway node between Cluster C2 and Cluster C6.To overcome this problem a distributed gateway may be introduced. A Distributed Gateway (DG) is a pair of ordinary nodes which belongs to different cluster but there is a direct link between them as shown in the figure 3.

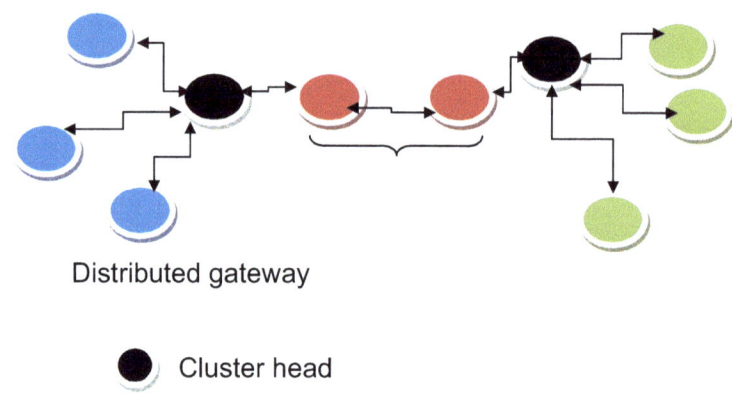

Distributed gateway

● Cluster head

Figure 3: Distributed Gateway

6.3 Clustering in MANET: Classification by properties, cluster-head capabilities, clustering process

According to A. Abbasi and M. Younis [15], clustering can be categorized as follows:

Table 1, Clustering properties:

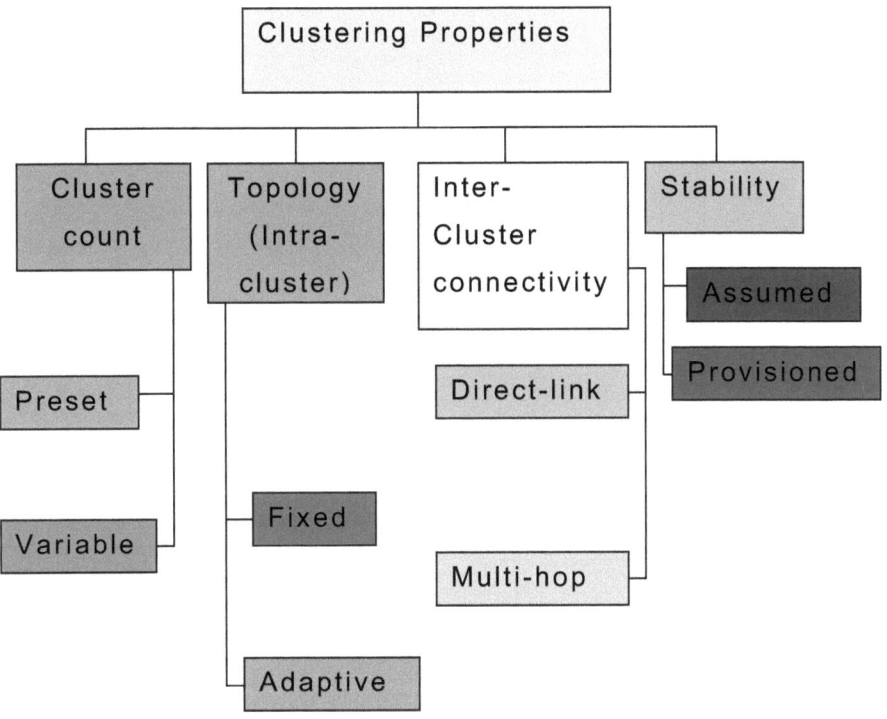

Table 2, Capabilities of cluster head

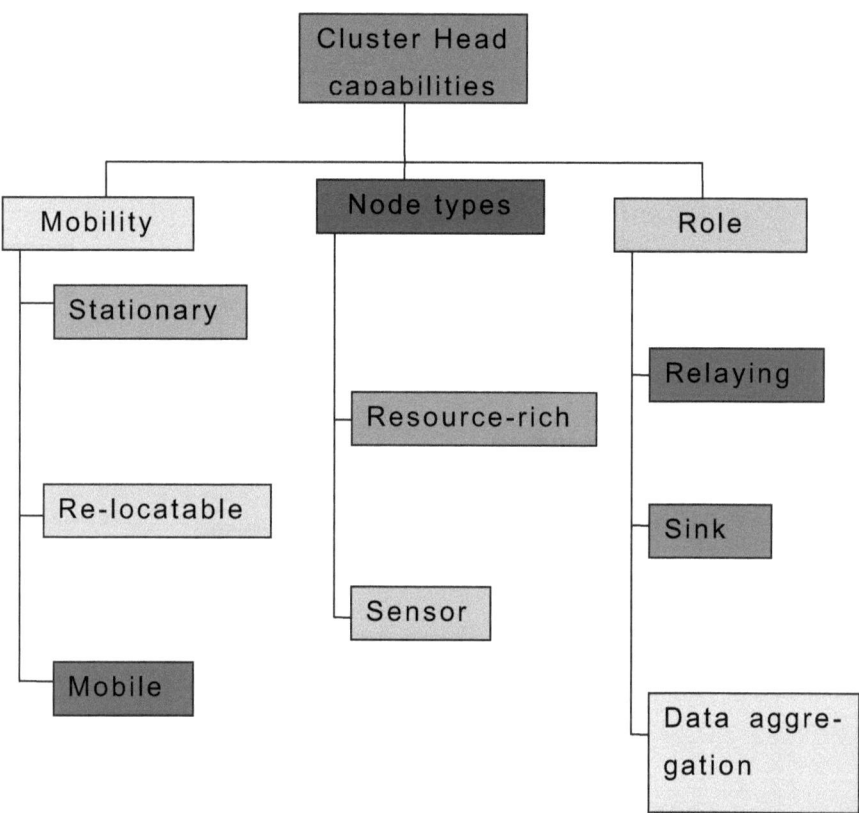

Table 3, Clustering process

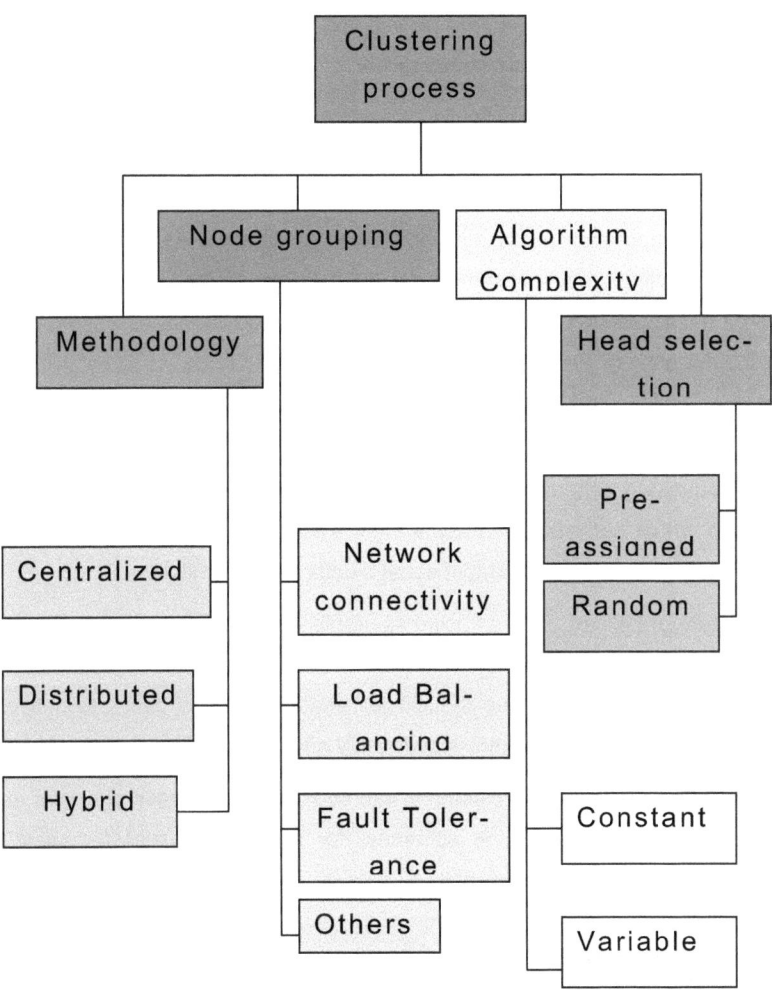

6.4 Cost of clustering

Clustering is important for a network to achieve scalability in the presence of a large number of mobile nodes and their mobility. However, a cluster-based MANET has some drawbacks because constructing and maintaining a cluster structure. Clustering usually requires additional computational cost, in contrast with a flat-based MANET.

The cost of clustering is a key issue to approve the scalability enhancement of a cluster structure. By analyzing the cost of a clustering scheme in MANET in different features of both qualitatively and quantitatively, drawbacks can be clearly specified. The clustering cost terms are stated below.

To maintain a cluster structure in a dynamically changing environment often requires sending and receiving of explicit messages between mobile nodes. When the underlying network topology changes rapidly and involves many mobile nodes, the clustering-related information exchange increases significantly. As a result, frequent information exchange may consume considerable bandwidth and drain mobile nodes' energy quickly. As an effect, upper-layer applications cannot be implemented due to the insufficiency of available resources.

6.4.1 Ripple Effect

Some clustering schemes may cause the cluster structure to be completely rebuilt over the whole network when some local events take place, e.g. the movement or "die" of a mobile node, resulting in some cluster head re-election

(re-clustering) [16-18]. This is known as 'ripple effect' of re-clustering. Ripple effect of re-clustering specifies that the re-election of one cluster head may affect the structure of many clusters and arouse the cluster head re-election over the network. Thus, ripple effect of re-clustering may significantly affect the performance of MANET.

Most schemes separate the clustering into two stages, formation of cluster and cluster maintenance. In some of the schemes, for the initial cluster formation of these schemes, a mobile node can make a decision to become a cluster head only after it exchanges some specific information of its neighbors and assures that it holds some specific quality in its neighborhood. With a cold period of motion, each mobile node can obtain precise information from neighboring nodes, and the initial cluster structure can be formed with some specific characteristics.

However, this assumption may not be applicable in an actual scenario, where mobile nodes may move randomly all the time in different directions.

6.4.2 Computation Round

It is an important metric. Computation round defines the number of rounds (turns) in which a cluster formation procedure can be completed. For clustering schemes relying on a frozen period of motion assumption, the computation round is an important metric (quality parameter) since the more number of rounds that a clustering algorithm needs for its cluster formation, the longer the frozen period that is required for mobile nodes.

In general, for most of the clustering algorithm, cluster formation procedure can be performed simultaneously in the whole network As a result, fast convergence in formation of a cluster is achieved. But in these algorithms, not all mobile nodes can settle on their status at the same time (say within first round), and they may need a non-constant number of turns to finish the initial cluster formation. The time required for these algorithms cannot be bounded and may vary for different topological changes in MANET.

Hence, the necessary exchange of explicit control messages, the ripple effect of re-clustering, and the stationary assumption for cluster formation are the key features or costs of a complex cluster-based MANET compared with a MANET with a simple flat structure.

Table 4: Clustering costs and their significance

Clustering cost	Significance
Ripple Effect	Re-election of a cluster head in MANET may affect the cluster structure.
Computation round (constant)	Number of turns or rounds that a cluster formation procedure can be completed.
Message passing complexity	Total amount of clustering related messages exchanged for cluster formation.
Explicit control of message	Between node pairs, clustering needs explicit clustering–related information exchanged.
Stationary assumption (cluster formation)	Mobile nodes to be considered static, in cluster formation phase.

6.5 Design parameters of clustering

6.5.1 Trust value

Measures the degree of trust i.e. how much any node in the mobile ad hoc network is trusted by its neighborhood nodes. In practice, its computed by taking average of trust values received from each neighboring nodes.

For computation, we consider that every mobile node has same intrusion detection (ID) mechanism to determine if a node is considered as trust or not by periodically collecting information about the behavior of each neighbor.

$$T = \frac{\sum_{i=0}^{N} T_i}{N} \qquad (1)$$

T_i is the received trust value from i^{th} node.

6.5.2 Degree

The degree of a node is defined as the number of neighbors of a given node, within a given radius. This parameter helps to decide for cluster-head selection, as node having higher value chosen to be a cluster-head.

6.5.3 Battery Power

This factor is the capability of a node to serve as long as possible. Since cluster-head has extra accountability and it must communicate as far as possible. Therefore, it must be the most powered node in the MANET.

6.5.4 The Max (maximum) value:

This parameter is used in the election procedure to elect a particular node as a cluster-head, from many nodes; the node which has a capability of handling the maximum number of nodes.

6.5.5 Stability (also defined as Mobility)

In order to avoid frequent roaming, the most stable node is elected as cluster-head. This stability may be computed using following metrics:

1. Distance:

The distance between two nodes A,B ($D_{A,B}$) is the number of hops between them, which can be obtained from the packets sent from one to other. To measure the distance a token message used in routing protocols. The likelihood of obtaining the number of hop counts between two nodes is clear and simple within all existed routing protocols.

2. The mean distance (MD):

Mean distance is measured as the average of distances between a node and all its neighbors.

3. Stability:

Defined as the difference between two measures of MD at t and t-1, it becomes large when the node goes far from its neighbors or whenever its neighbors are going in other direction than the one taken by the considered node. This value is compared with D and a node is considered as most stable if it has the less value of ST.

$$ST_A = MD_t - MD_{t-1}$$

6.6 How to maintain Structure of Cluster

6.6.1 Sending beaconing signal:

In order to maintain the structure of clusters, cluster head broadcast periodically token messages called beacons. These beacons contain commands sent to cluster-members to collaborate in order to execute any one of the cluster management commands. Beacon has two fields; the first one contains the certificate of the cluster-head, followed by the code of the command.

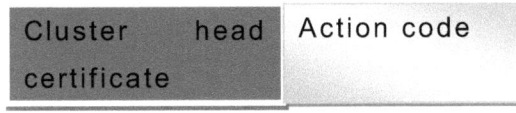

Figure 4: Beacon Structure

Beacons are periodically broadcast in the network, using flooding technique. For a large number of nodes, flooding suffers from congestion of medium, redundant packets, probability of collision and security risks.

Chapter 7:
Proposed Algorithm

The proposed energy based clustering algorithm is described as following steps:

Step1: Initially few nodes are selected as co-ordinator nodes in network.

Step2: Decide the neighbors of the co-coordinator node such that nodes are within the defined transmission range (T_r) of the co-ordinator node.

Step3: If the nodes are within the Transmission Range (T_r) calculate mobility. The mobility is calculated as considering x,y and z axis.

$$\text{Mobility } (M) = \frac{1}{T} \sum_{t=1}^{T} \left(\sqrt[2]{(Xt - Xt + 1)2 + (Yt - Yt + 1)2 + (Zt - Zt + 1)} \right)$$

Where (X_t, Y_t, Z_t) and (X_{t+1}, Y_{t+1}, Z_{t+1}) are the co-ordinates of the node at time t and t+1 respectively.

Step4: If the mobility of the node is higher than its threshold value, then the network is unstable, then the co-ordinator waits for some delay (D).

Step5: The co-ordinator assigns each node an unique ID randomly.

(Now the co-ordinator and as well as nodes periodically broadcast a 'Hello' message and each node maintain a neighbor list.)

Step6: Now the Cluster Head Selection Factor is computed, defined as:

CSF = W1*E +W2*C + W3*I

Where E is the Energy,

C is the connectivity,

I is the node ID

(The values of these three factors are mapped to some normalized values and the weights can be assigned according to their relative importance. Here we give more importance to the Energy of a node).

Step7: The computed CSF is used to calculate the delay for a node to announce as Cluster Head. The higher the CDF the sooner the node will transmit.

Step8: Upon receiving a message a node first verifies whether the node ID and the Cluster ID in the received message are same.

(Same node and Cluster ID means that message has been transmitted from a cluster head.)

Step9: If the receiving node does not belong to any cluster and the received CDF is better than its own score two cases may occur:

Case 1: If the current node receiving a better scored message is not a Cluster Head itself, as an ordinary node it can immediately mark down the best Cluster and wait until schedule announcement. This node will stay in its committed cluster after its announcement.

Case 2: If the current node is a cluster head receiving better CDF that this may need to switch better cluster.

Chapter 8:
Clustering Simulation Framework for MANET

8.1 Architectural view of Network Simulator

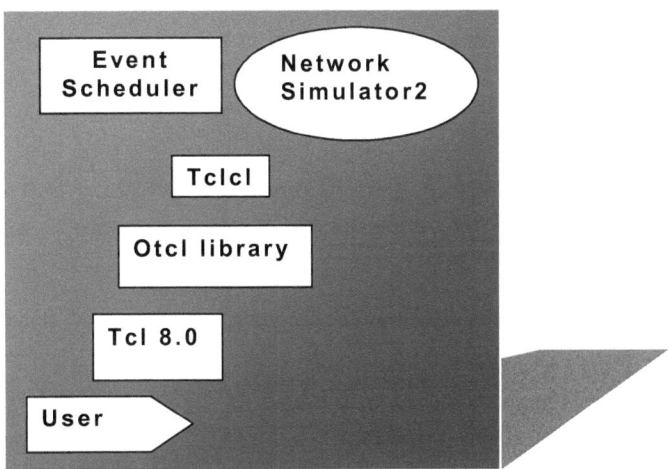

Figure 5: Architectural view

As in fig. 14, a user sees of standing at the left bottom corner, designing and running simulations in Tool command language (Tcl) using the simulator objects in the Object Tool Command Language (OTcl) library. The event scheduler and network components are implemented in C++. These are available to OTcl through an OTcl linkage. OTcl is implemented using TCL with Classes (tclcl). NS is an extended Tcl interpreter with network simulator libraries. When a simulation is completed, Network simulator creates text-based output files that contain detailed simulation data in the input OTcl or Tcl script. The data can be used for simulation analysis or as an input to Network Animator (NAM), a graphical simulation display tool. It can graphically point out information, for example throughput, number of packet drops at each link etc.

8.2 Clustering Framework

Clustering Framework is a library, used for developing and comparing clustering algorithms in ns2.

8.2.1 Concepts

In NS2, a clustering algorithm is implemented in a layer placed between the MAC and Link layers in the mobile node structure. For a given topology, the clustering algorithm creates an overlay over the network visibility graph. The MAC layer used by simulations scripts is IEEE 802.11. This is a CMU extension of NS2 distribution. MAC layer provides mechanism to send messages both in the form of unicast and broadcast messages.

Clustering algorithms proceed in steps:

An algorithm can have need of nodes to swap messages to create a first rich backbone network. It tries to reduce the number of backbone nodes by exploiting some topological information.

Clustering Sub layer: Each algorithm in clustering, which are distributed in nature, is called Clustering Sub Layer (CSL).

8.2.2 Steps of simulated algorithm

1. When simulation starts the first CSL of each node in the network receive a message called 'start Modulo'.

2. The start procedure launches the clustering procedure concurrently at each node in the network. Please note that, the execution time of each algorithm is the same simulation time for each node.

3. Each Clustering Sub Layer exchanges messages with neighbor nodes in the MANET using primitives provided by the MAC layer.

4. When a CSL installed in a node finishes its course of action (when for example the node contacted every neighbor to exchange some useful information) it calls the module known as 'method end Module'.

5. All information collected by the dumper of the current clustering layer is put on the screen as output.

6. If the current algorithm is the last one, the simulation ends. In that case, the global statistics are collected and are printed out.

If not, the simulator restarts the execution from step (1) by using the next Cluster Sub Layer that defines the clustering algorithm.

Clustering Sub Layer can dump its own results on standard output and on the global result collector. This is a class that reports a collection of global metrics. Each CSL can put connectivity graph, along with the nodal energy consumption expressed in terms of packets transmitted and received. It can also provide information in terms of the energy consumed. It is also possible to dynamically compile clustering algorithms by mean of the combination of one or more CSL. The template directory contains some Tcl files that are used in the implementation clustering solutions. Each layer dump follows its own 'preamble'.

8.3 Simulation Result

In our analysis for a single hop cluster we have shown the statistics of packet transfer from a node to an external node through the gateway controlled by a cluster head.

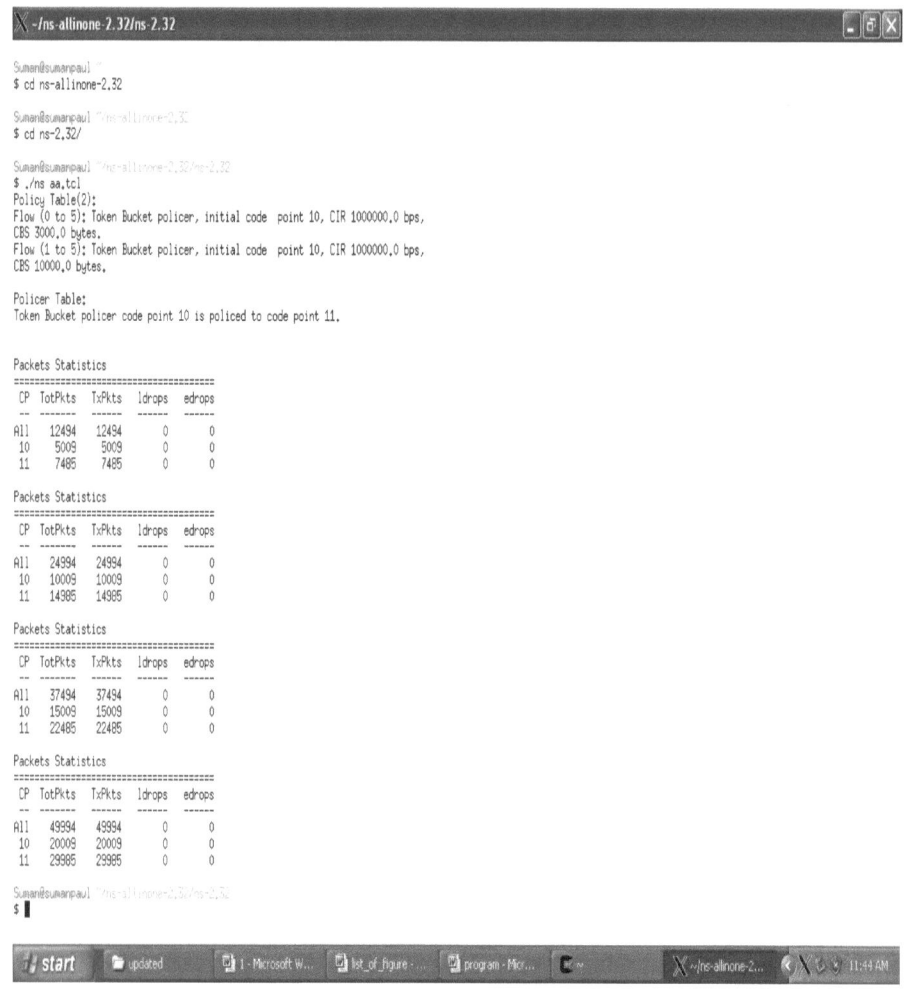

Figure 6: Snapshot of output

Further Reading

[1] Ching -Chuan Chiang, Hsiao-Kunag Wu, Winston Liu and Mario Gerla, "Routing in Clustered Multihop, Mobile Wireless Networks with Fading Channel," IEEE Singapore International Conference on Networks, SICON'97, pp. 197-211, Singapore, 16.-17. April 1997, IEEE

[2] Connectivity based k-hop Clustering in in Wireless Network -FABIAN GARCIA NOCETTI and JULIO SOLANO GONZALEZ, Telecommunication Systems 22:1–4, 205–220, 2003

[3] Secured Clustering Algorithm for Mobile Ad Hoc Networks, B. Kadri, A. M'hamed, M. Feham, IJCSNS International Journal of Computer Science and Network Security, VOL.7 No.3, March 2007

[4] M. Bechler, H.-J. Hof, D. Kraft, F. Pählke, L. Wolf. A Cluster-Based Security Architecture for Ad Hoc Networks. IEEE, INFOCOM 2004.

[5] S.Basagni "distributed Clustering for Ad Hoc Network",in International Symposium on parallel Architecture , algorithns,and Networks.

[6] Effect of mobility models on Clustering in MANET –S chinara, S.K. Rath., National Conference on Smart Communication Technologies and Industrial Informatics.

[7] Mobility-based d-Hop Clustering Algorithm for Mobile Ad Hoc Networks, Inn Inn ER , Winston K.G. Seah2,1-WCNC 2004 / IEEE Communications Society.

[8] A. B. McDonald and T. F. Znati. A mobility-based framework for adaptive clustering in wireless ad hoc networks. IEEE Journal on Selected Areas in Communications, 17(8):1466-1486, Aug. 1999.

[9] A Low-Maintenance Energy-Aware Clustering Algorithm for Wireless Ad-hoc Networks - F. Foroozan, S. Datta, 1-4244-0495-9/06,IEEE.

[10] C.-C. Chiang, H.-K. Wu, W. Liu, and M. Gerla,"Routing in Clustered multihop Mobile Wireless Networks with Fading Channel", Proceedings of IEEE Singapore International Conference on Networks (SICON'97).

[11] An Huiyao, Lu Xicheng, and Peng Wei. A Cluster-Based Multipath Routing for MANET. Med-Hoc-Net 2004, The Third Annual Mediterranean Ad Hoc Networking Workshop.

[12] Haas, Z.J., Pearlman, M.R. and Samar, P. Zone, Routing A Scalable Key Management and Clustering Scheme for Ad Hoc Networks - Jason H. Li, Renato Levy and Miao Yu, Bobby Bhattacharjee, INFOSCALE '06. Proceedings of the First International Conference on Scalable Information Systems, May 29-June 1 2006, Hong Kong.

[13] A Mobility-Based Framework for Adaptive Clustering in Wireless Ad Hoc Networks - A. Bruce McDonald , Taieb F. Znati -IEEE JOURNAL ON SELECTED AREAS IN COMMUNICATIONS, VOL. 17, NO. 8, AUGUST 1999.

[14] J.J. Garcia-Luna and M. Spohn, "Source Tree Adaptive Routing Internet Draft," draft-ietf-manet-star-00.txt, work in progress, October 1999. / J.J. Garcia-Luna. M Spohn, "Source-Tree Routing in Wireless Networks," Proceedings of the 7th International Conference on Network Protocols, IEEE ICNP 99, Toronto, Canada, pp. 273-282, IEEE, October 1999.

[15] A. Abbasi and M. Younis, "A survey on clustering algorithms for wireless sensor networks," *Comput. Commun.*, vol. 30, nos. 14–15, pp. 2826–2841, Oct. 2007.

[16] A. D. Amis and R. Prakash, "Load-Balancing Clusters in Wireless Ad Hoc Networks," in *Proc. 3rd IEEE ASSET'00*, Mar. 2000, pp. 25–32.

[17] T. J. Kwon *et al.*, "Efficient Flooding with Passive Clustering - an Overhead-Free Selective Forward Mechanism for Ad Hoc/Sensor Networks," in *Proc. IEEE*, vol. 91, no. 8, Aug. 2003, pp. 1210–20.

[18] M. Chatterjee, S. K. Das, and D. Turgut, "An On-Demand Weighted Clustering Algorithm (WCA) for Ad hoc Networks," in *Proc. IEEE Globecom'00*, 2000, pp. 1697–701.